WRITERS REPUBLIC

What Does an Owl Do?

Math Book

Andrea Davis

WRITERS REPUBLIC L.L.C.
515 Summit Ave. Unit R1
Union City, NJ 07087, USA

Website: *www.writersrepublic.com*
Hotline: *1-877-656-6838*
Email: *info@writersrepublic.com*

Ordering Information:
Quantity sales. Special discounts are available on quantity purchases by corporations, associations, and others. For details, contact the publisher at the address above.

Library of Congress Control Number: 2024903180
ISBN-13:   979-8-89100-427-6   [Paperback Edition]
                979-8-89100-426-9   [Digital Edition]

Rev. date: 10/23/2024

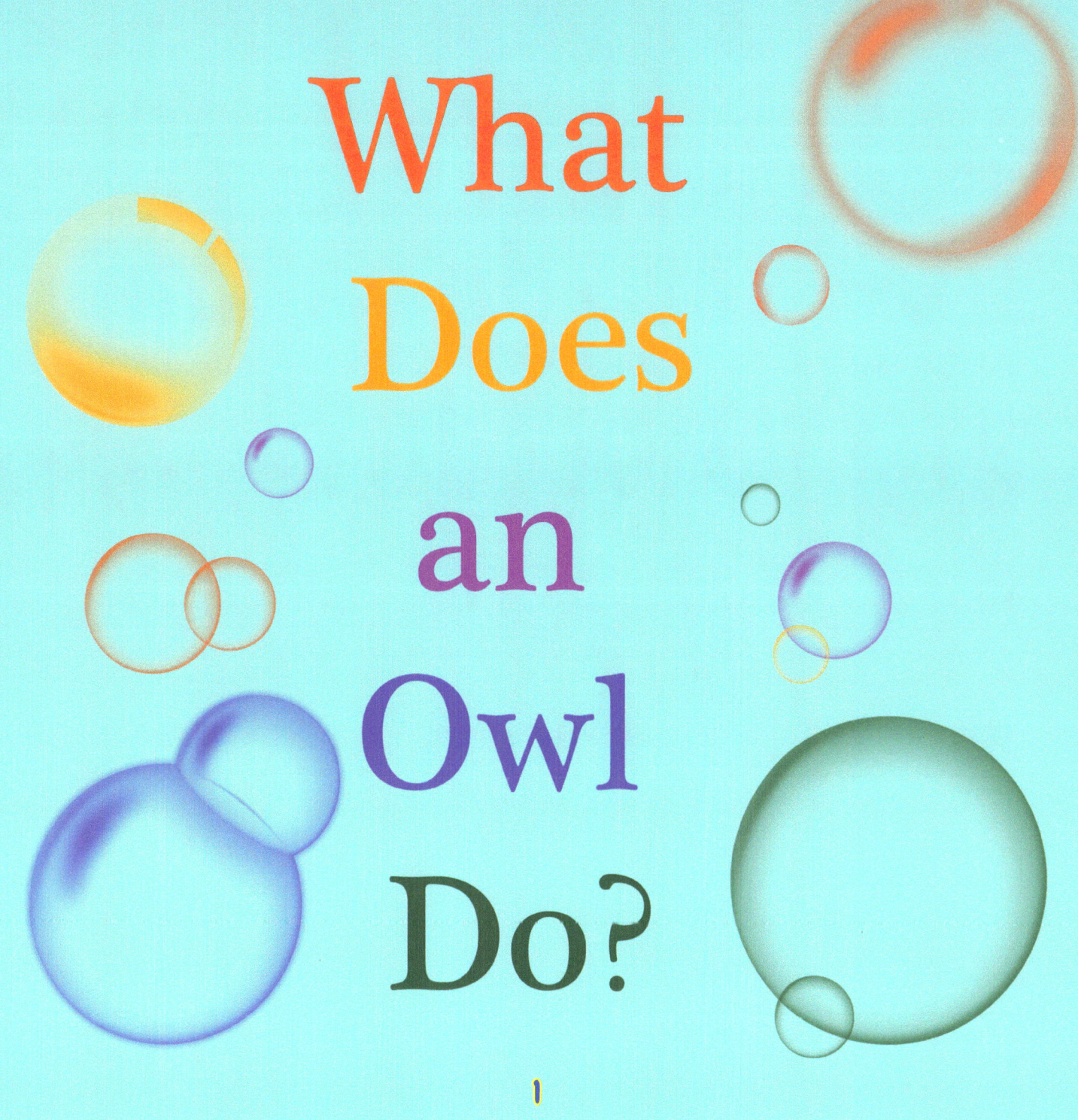

What Does an Owl Do?

What does an Owl do?

They can fly

What does an owl do?

They make a sound woo woo on a tree bark.

What does an Owl do?

They live in a tree type is called cottonwood where they set on the top of the branch.

What does an owl do?

They can turn their neck to neck.

How many owls are there?

They have 250 species of birds.

The end.

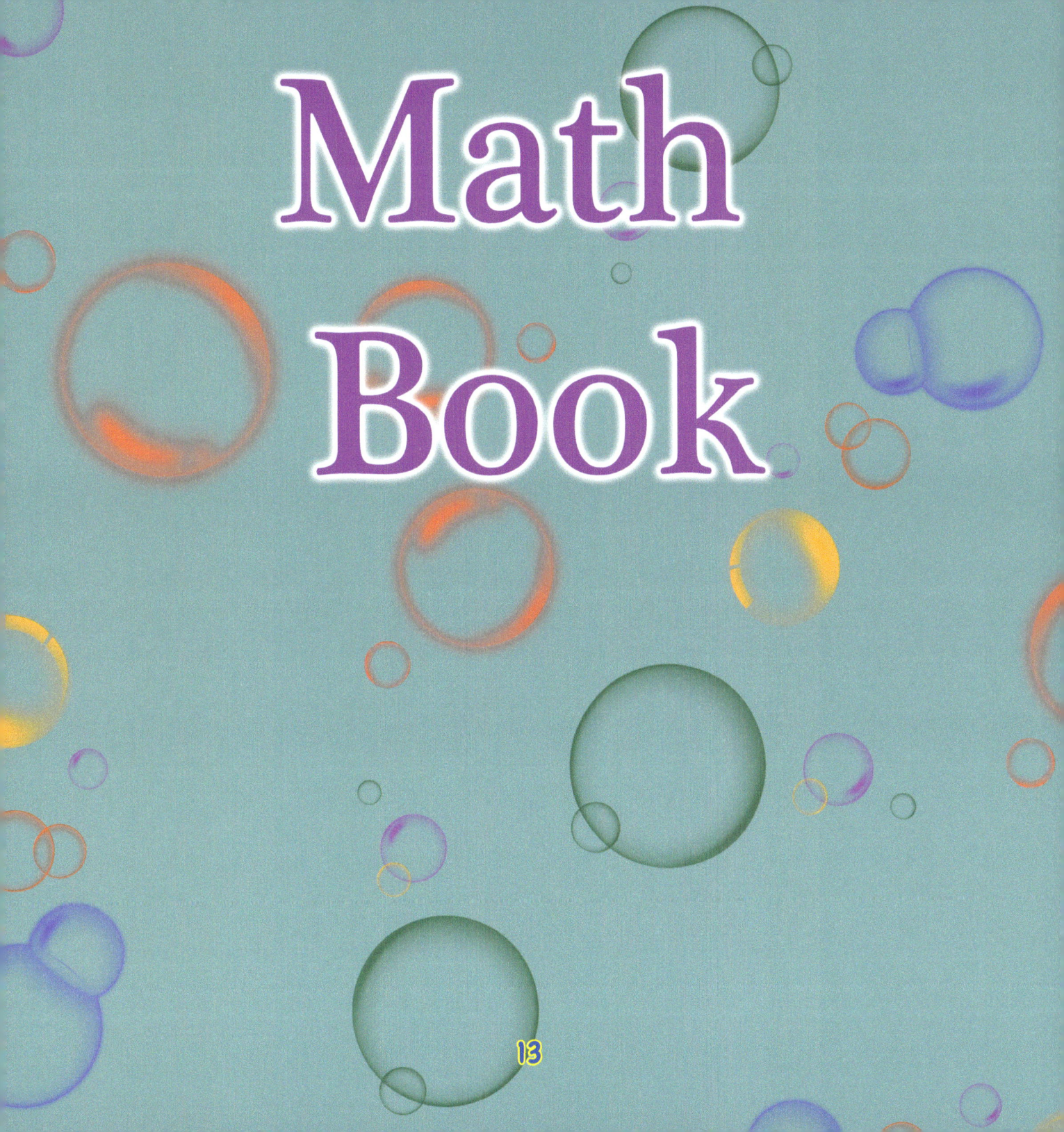

Math
Book

13

1 + 1 = 2

Math

$1+2=3$

$$1 + 4 = 5$$

1 + 5 = 6

The end.